AUTUMN
IN THE WOODS

written and photographed
by
Mia Coulton

It is autumn.

It is fun to walk in the woods and look for signs of animals getting ready for winter.

Look down on the ground.

Look at all the **acorns**.

Acorns and other nuts are food for many animals getting ready for winter.

Look at the big hole in the tree.

Look inside the hole.

This tree will make a good winter home for an animal.

Look at the log on the ground.

It is hollow.

Maybe a little mouse

will make this its winter home.

Look at the little, brown **woolly bear** on the leaf.

The woolly bear will make its winter home under the leaves on the ground.

Look up in the tree.

It is a young **bald** **eagle**.

Eagles fly away to warmer places to look for food if the lakes and streams nearby freeze over.

Look all around the woods in the autumn.

The animals are getting ready for winter.

Glossary

acorns — The nuts of an oak tree

autumn — The season that comes after summer and before winter, sometimes called *fall* for the falling leaves during this season

bald eagle — A large, brown North American bird whose head feathers turn white around age five

hollow — An empty space inside something

woolly bear — A hairy, brown and black caterpillar